Effects of Alcoholic Excess on Character

Milner J. Fothergill
Henry S. William
T. D. Crothers

Effects of Alcoholic Excess on Character

LM Publishers

Effects of Alcoholic Excess on Character[1]

A great deal of attention has of late years been bestowed upon the subject of alcoholic indulgence. The importance of the subject warrants this, and even calls for still further attention. There are differences of opinion as to the use of alcohol; there are comparatively none as to the abuse of it. Leaving then, for the present, the question of the use of alcohol in disease, its effect upon the body temperature, and its position as a food, we may profitably engage ourselves for a little time with its social effects, alike upon the individual and the masses, especially in reference to its influence upon the mental manifestations of brain-activity. It is now universally acknowledged that mental alterations follow physical

---

[1] Milner J. Fothergill

modifications of the brain, as seen in the various forms of insanity. We well know how profound is the influence exerted by alcoholic excess upon the brain, and through it upon the character. Unfortunately, the effects of continued alcoholic excess are but too frequently forced upon our attention.

The most pronounced product is found in the hopeless drunkard, who, in squalid rags, with rotten tissues, the embodiment of intellectual and moral degradation, utterly beyond hope, the line of possible restoration long past, hangs around the tavern-door, and, with the odor of alcohol floating on his breast, whiningly begs a copper from the mass of vitality around him, of which he himself is a withered and decaying branch. This man is incapable of labor; he is unwilling to entertain the idea of toil. He is beyond any capacity for labor; he is no longer capable of discharging his duty as a citizen; he is a social parasite of the lowest and foulest order, as useless as a tapeworm. He has abandoned all self-respect, because there is nothing left in him for himself or anyone else to respect. He is a shameless liar, who will make the most solemn protestations as to the truth of what it is patent enough is false. There is no depth of moral degradation to which he will not descend for the means to purchase a little more of the fluid which has ever been his bane.

Betwixt him, however, and his patrons, many of whom enter the tavern to celebrate some little matter by a glass together, there is a potential association, not always at first sight readily apparent. The effect of alcoholic indulgence is seductive; and it often creeps on unobserved, doing much irretrievable mischief ere its presence is unmistakable. It is not the intention of the writer here to discuss the question of the moderate use of alcoholic beverages, but rather to point out the fruits, the evil consequences, of excess. Betwixt the hopeless drunkard and the casual taker of a social glass there are a thousand grades and modifications. Nor does it necessarily follow that the one shall degenerate into the other; very commonly he does not; but, unfortunately, he may, and not unfrequently does. Too frequently, indeed, the practice grows, especially in those who naturally lack self restraint, or cannot control their impulses, however capable in other respects. The dangers of alcoholic allurement are various in their degrees of potency in different individuals.

Not only that, but there is no little influence exercised by the immediate motives for which alcohol is taken. The future progress of the individual indulging in alcoholic excess is widely different, according to the mental attitude at the time. Thus, betwixt the man who has been taking alcohol to excess at intervals extending over many years, and the young woman who is just commencing to drink because she is unhappy, there is a wide gulf. The one, so far as the alcohol is concerned, will probably live to an advanced age; the prospects of life in the other are very poor, and the ruin will be swift and complete. In the one there are long intervals of sobriety, during which the effects of the debauch will, to a great extent, wear off; in the other the act will be repeated as often as opportunity will permit; one act of indulgence will lead to, indeed will induce another, and the oft and quickly repeated act will become a constant habit, whose effects are soon felt. It is not in women alone that the hopeless nature of drinking habits in certain susceptible organisms is manifested; it is equally seen in men where the nervous system lacks stability.

The deceptiveness, the utter untrustworthiness, the subtle craft, the falsehood, which women of culture even will develop under the influence of alcoholic cravings, have shocked many persons. The habitual drunkard, however produced, always exhibits these characteristic signs of moral degradation. The deterioration of character produced by protracted drunkenness is notorious. While the intellect becomes enfeebled by excess, the moral character becomes profoundly modified; the forces which ordinarily restrain others are in abeyance— perhaps too often their influence has gone forever; the indifference toward the interests of others progresses alongside a waxing selfishness, a complete absorption in self. So long as they can procure what they themselves crave for, confirmed drunkards are indifferent as to how others may suffer for, or be injured by, their selfishness. The ordinary feelings of parent or husband are too often overruled by the consuming passion; the wonted consideration for those who used to be dear to them has given way to an inordinate egotism. Not uncommonly, indeed, there is developed a vein of devilish mischievousness which delights in injuring those whom they ought to protect—a sort of malice, closely resembling the viciousness of certain animals. Of course, all drunkards are not exactly alike; the ruin still preserves the general outline of the primitive structure.

These statements may seem to some to be unnecessary as being already too well known, and too notorious to need any reference to them. But it is just because they are so well known and so indisputable that they are adduced here. Having thus laid firmly down the well-marked consequences of persistent alcoholic excess, it is possible to proceed to consider the less pronounced conditions, and to trace the course of the downward progress. It is evident that there must be many intermediate stages betwixt the commencement and the end of such a course—that some of the deteriorating effects of alcohol must be experienced long before the final stage is reached.

It may be well to speak in general terms of the indication of this direction, of this retrograding and degenerative process. The best subjects for the study of the social effects of alcoholic excess are furnished by the humbler classes: firstly, because the effects are more palpable among them with their limited resources, where excess in one direction means deprivation in another; and, secondly, because they present fewer complications, fewer elements of error to be encountered, than is the case in the more complex condition of affluence. It must not, however, for one moment be assumed that the evil consequences of alcoholic excess are confined to the humbler classes. No position in life will secure the individual against the unpleasant consequences of such self-indulgence, or prevent his reaping as he has sown. A momentary digression may be permissible at this point; it will help to illustrate what is to be said shortly. In an elaborate paper on alcohol, read before the Medical Society of London last winter, by Dr. Lauder Brunton, F. R. S., to which was awarded the society's medal, it was stated that the first effects of alcohol are felt in the higher or controlling portions of the brain. The consequences are, that the lower or animal impulses manifest themselves, freed from the control to which they are ordinarily subject. Such are the first symptoms of intoxication, after the stage of exhilaration has been passed.

Then the motor centres are implicated; and complex movements, like walking, ordinarily habitual, require a conscious effort for their execution, and even then the performance is imperfect. After this the lower portions at the base of the brain are involved, leaving nothing but the respiration and the circulation in action; while still further intoxication arrests these movements, and the organism perishes.

It is this effect upon the higher centres of the brain which induces the most disastrous social outcomes of alcoholic indulgence. The person who takes alcohol to excess becomes a lower form of being by comparison with those around him. This is seen alike in the individual and in the aggregate. There is a diminution so brought about in the power to exercise self-control, and to estimate aright the claims of the future upon the present; there is produced a state of thriftlessness and recklessness, and a lack of consideration for others. These effects are demonstrated distinctly in two out of many practices. The one is that form of improvident self-indulgence—early and premature marriage, where the rite is robbed of all sacredness, and degraded to a mere form of license for unrestrained indulgence. This is common in the pit districts of Durham, where comparative children present themselves to be married, who can scarcely have realized the gravity of the step they are taking. Another form, sadly too common, is that of living in any hovel where the rent is small, so that a larger sum weekly remains over to be spent in drink. This is especially found amid the Irish in towns. The moral effects of decent houses, with sufficient sleeping accommodation for the two sexes, are well known; the disastrous immoral effects of huddling different sexes and ages together, from want of proper sleeping-space, are equally notorious. Not only are these evil consequences of such overcrowding produced, but, when alcohol gives the rein to the passions, these consequences are aggravated and intensified. There are then blended the direct and the indirect outcomes of alcoholic excess, each of which aggravates and adds to the other.

It is unnecessary to multiply illustrations of the deterioration of character induced by alcoholic indulgence. Just one more may be adduced. When recently on a visit to the South Yorkshire Asylum, near Sheffield, Dr. Mitchell, its accomplished superintendent, informed me that even in the victims of mania and general paralysis there was a marked contrast in the degree of violence manifested betwixt the patients arriving from the purely agricultural districts and those from the iron-work and colliery districts. This he attributed partly to their rougher ways and to the nature of their occupations, but still more to the drinking habits of the latter. That these last formed the chief factor in the production of the result was rendered probable by the large proportion of female subjects of general paralysis in that asylum. This disease is comparatively rare among women; and its prevalence among females from these districts Dr. Mitchell attributed to the habits of the women being allied to those of the men, especially as regards indulgence in drink.

Such, then, are some of the grave social outcomes of systematic indulgence in alcohol which arrest our attention. We have seen that its effects upon the nervous system are such as to give the rein to the lower centres, chiefly by lessening the control exercised by the higher and restraining portions of the brain. Man escapes from his wonted self-restraint when under the influence of alcohol, and stands before us with his fundamental character revealed. The groundwork of his character is exposed by the removal of the demeanor which he has carefully cultivated. The outside cover is withdrawn; all, or nearly all, that self-education or cultivation has given, is temporarily taken away. Through the revelations so made by alcohol we not rarely find that even in staid and proper men the tiger and the ape have not entirely died out. The animal propensities are thus discovered to have been concealed rather than subdued. For the time being the intoxicated individual is reft of much that not only he but his ancestors for generations back have studiously cultivated. For the time being he is a lower type of man. About the truth of this statement there can be no doubt.

The progress of physiological psychology, of the investigation of the workings of the mind, has taught us, in unmistakable accents, the strong tendency which exists for a habit to be formed by repetition of anything. After a thing has been done several times it becomes exceedingly easy to do it again. There is, in fact, in the nervous system a great readiness to take on an attitude which has been assumed before. We all recognize how it becomes necessary for everyone to rehearse a part before acting it, and how quickly a species of habit or imitative attitude is formed. It is widely recognized that practice makes perfect, and that what was once difficult becomes easy by repetition of it. These are but illustrations of a law universally acknowledged. We all know how important it is to avoid what may become a habit. Consequently, we can see distinctly and with painful clearness that repeated indulgence in alcoholic stimulation, not necessarily extending to visible intoxication, must tend, by virtue of this law, to modify and mould the character. Under alcohol the individual becomes sanguine, reckless, careless of consequences, boastful, and indisposed to sober calculation; he also becomes self-assertive, arrogant, and boisterous; there exist a certain impulsiveness and impatience of control, and a distinct tendency to reach certain ends by violence, if other measures do not seem likely to be successful. In fact, we see the habitual self-restraint slowly developed by the exertions of many ancestors, by the efforts of the individual himself, aided by the training given by education, for the time being withdrawn to a great extent. Every time this act is repeated the greater the tendency for the character to manifest its lower rather than its higher forms. The character is indeed being slowly modified, and that, too, in a most undesirable direction. It is being gradually deprived of much that a slow process of civilization has given it.

It must be patent to all that the direction just depicted is that in which our town populations are distinctly moving. I have sketched elsewhere (the *Alliance News,* July 22 and 29, 1876) those circumstances in the condition of the masses which, in my opinion, render something more than the mere removal of temptation necessary in the practical treatment of intemperance. Whatever may be the effect of monotonous occupations, of bad hygienic surroundings, of improper food in infancy, and the physical deterioration which results from large aggregations of individuals, there can be no question but that repeated alcoholic indulgence is gradually modifying the character of the masses. It is seen in the growing insubordination, in the turbulence, the impatience of control, in the tendency to assert their opinion on subjects of which they are not in a position to judge; it is demonstrated in the growing thriftlessness, in the marked inclination to increase the number of their holidays, in the spending of a large portion of their weekly earnings on Saturdays and Sundays, till the middle of the week is a time of comparative destitution. It is illustrated by the columns of our daily press in the increased acts of violence perpetrated under the influence of alcohol by men accustomed to be intoxicated; and by the increase of disease of the nervous system, especially excitable forms of mania, found most commonly amid the population of certain industrial districts. Doubtless there are many thrifty, sober, self-respecting, and industrious working-men; but it cannot be denied that the proportion of such men to the whole body is less now than it was in days not far gone by. The character of the masses has been undergoing grave modifications in recent times; and the fashioning hand of alcohol can be clearly traced in the production of the results.

There is, however, a still grimmer aspect of this subject than even the effects of repeated indulgence upon the mental attitude of the individual himself, and that is the influence which such indulgence will and must exercise upon the nervous system of his offspring. We are all of us "the outcome of the coöperation of countless ancestral forces," and each of us is individually indebted to his ancestors for every act of self-restraint, every act of self-denial or control exercised by each of them. Our forefathers, in forming their own character, were insensibly fashioning ours. "The fathers have eaten sour grapes, and the children's teeth are set on edge," is sternly true. What, then, can we legitimately expect to be manifested in the next generation? Further, if children are taught to frequent taverns, and drink there on holidays and Sundays, by their fond but foolish parents, what effect must this exert upon the character during the plastic period of youth and growth? How far are the inherited mental constitution and nervous system, already depraved to start with, still further modified by such experience of the individual in childhood, when "wax to receive but marble to retain?" Conditions already hard enough upon the child are aggravated by indiscretions perpetrated toward the infant before its own free will and choice can be called into play, before it is responsible for its own actions. Not only have its parents given it an imperfect organization, but they are prejudicing its chances of self-evolution before it has had an opportunity of forming its own decision—it is handicapped alike by descent and by mischievous early training.

The habit of frequenting taverns, of drinking, and of feeling the self-satisfaction so induced, leads to still further indulgence in alcohol by half-grown youths; and so the inherited character is still further deteriorated. The increasing loss of self-control leaves such beings less and less capable of resisting the temptations, the allurements of the public-house. The impulsive and less perfectly controlled nervous system craves more and more for the alcoholic stimulant; and the longings are intensified accordingly. The repeated visits to the tavern grow into a custom, and what commences as an irregular practice becomes crystallized into a habit.

Nor is it in youths alone that the drinking customs of the day are seen in their evil and sombre aspects. The number of respectable girls seen now at public bars is a contrast to what obtained but a few years ago. Up to a recent period, if a girl were known to frequent taverns, her character was gone; and it was rarely that a well-conducted girl was seen in a public-house, and then only with her sweetheart or some male relatives. But now it is sadly different. From familiarity with bars as an outcome of excursions, and even more from the associations of the music-hall, girls, capable of better things, are not now apparently conscious of any impropriety in being in a public-house without male friends; and the painful spectacle of seeing young girls under twenty treating each other at a public bar is a sadly too common occurrence. How can a girl, with the mobile nervous system of her sex, be fitted to be a mother, and to counteract the evil tendencies of alcoholic indulgence in the father, if she herself have been subjected to the same influence? With the facts of inheritance before us, what may we expect, what must we apprehend as to the condition — the future prospects — of the generation following immediately after this one? As our forefathers insensibly and unconsciously built up the character of the present generation, so it, in its turn, is fashioning the character of its successors, its unborn offspring. No wonder, then, that the *morale* as well as the *physique* of the masses in large towns is undergoing already retrograde changes; and that the present condition fills the minds of observers of social progress with gloomy forebodings as to the future. The progress of civilization has endowed us with a measure of self-control, has tended to subordinate the unit of the mass—to encourage the evolution of the citizen as compared to the mere individual. The effect of alcoholic indulgence to excess is to institute retrogressive changes, and to undo, to a great extent, what civilization has slowly achieved.

Alcohol and the Individual[2]

Some very puzzling differences of opinion
about the use of alcoholic beverages find
expression. This is natural enough, since
alcohol is a very curious drug, and the human
organism a very complex mechanism. The
effects of this drug upon this mechanism are
often very mystifying. Not many persons are
competent to analyze these effects in their
totality. Still fewer can examine any of them
quite without prejudice. But in recent years a
large number of scientific investigators have
attempted to substitute knowledge for
guesswork as to the effects of alcohol, through
the institution of definitive experiments. Some

---

[2] By Henry Smith William

have tested its effects on the digestive apparatus; others, its power over the heart and voluntary muscles; still others, its influence upon the brain. On the whole, the results of these experiments are singularly consistent. Undoubtedly they tend to upset a good many time-honored preconceptions. But they give better grounds for judgment as to what is the rational attitude toward alcohol than have hitherto been available.

The traditional rôle of alcohol is that of a stimulant. It has been supposed to stimulate digestion and assimilation; to stimulate the heart's action; to stimulate muscular activity and strength; to stimulate the mind. The new evidence seems to show that, in the final analysis, alcohol stimulates none of these activities; that its final effect is everywhere depressive and inhibitory (at any rate, as regards higher functions) rather than stimulative; that, in short, it is properly to be classed with the anesthetics and narcotics. The grounds for this view should be of interest to every user of alcohol; of interest, for that matter, to every citizen, considering that more than one thousand million gallons of alcoholic beverages are consumed in the United States each year.

I should like to present the new evidence far more fully than space will permit. I shall attempt, however, to describe some of the more significant observations and experiments in sufficient detail to enable the reader to draw his own conclusions. To make room for this, I must deal with other portions of the testimony in a very summary manner. As regards digestion, for example, I must be content to note that the experiments show that alcohol does indeed stimulate the flow of digestive fluids, but that it also tends to interfere with their normal action;

so that ordinarily one effect neutralizes the other. As regards the action on the heart, I shall merely state that the ultimate effect of alcohol is to depress, in large doses to paralyze, that organ. These, after all, are matters that concern the physician rather than the general reader.

The effect of alcohol on muscular activity has a larger measure of popular interest; indeed, it is a question of the utmost practicality. The experiments show that alcohol does not increase the capacity to do muscular work, but distinctly decreases it. Doubtless this seems at variance with many a man's observation of himself; but the explanation is found in the fact that alcohol blurs the judgment. As Voit remarks, it gives, not strength, but, at most, the feeling of strength. A man may think he is working faster and better under the influence of alcohol than he would otherwise do; but rigidly conducted experiments do not confirm this opinion. "Both science and the experience cf life," says Dr. John J. Abel, of Johns Hopkins University, "have exploded the pernicious theory that alcohol gives any persistent increase of muscular power. The disappearance of this universal error will greatly reduce the consumption of alcohol among laboring men. It is well understood by all who control large bodies of men engaged in physical labor, that alcohol and effective work are incompatible."

It is even questionable whether the energy derived from the oxidation of alcohol in the body can be directly used at all as a source of muscular energy. Such competent observers as Schumberg and Scheffer independently reached the conclusion that it cannot. Dr. Abel inclines to the same opinion. He suggests that "alcohol is not a food in the sense in which fats and carbohydrates are food; it should be defined as an easily oxidizable drug with numerous untoward effects which inevitably appear when a certain minimum dose is exceeded." He thinks that alcohol should be classed "with the more or less dangerous stimulants and narcotics, such as hasheesh, tobacco, etc., rather than with truly sustaining foodstuffs." Some of the grounds for this view will appear presently, as we now turn to examine the alleged stimulating effects of alcohol upon the mental processes.

Alcohol as a Brain Stimulant

The celebrated physicist Von Helmholtz, one of the foremost thinkers of the nineteenth century, declared that the very smallest quantity of alcohol served effectively, while its influence lasted, to banish from his mind all possibility of creative effort; all capacity to solve an abstruse problem. The result of recent experiments in the field of physiological psychology convince one that the same thing is true in some measure of every other mind capable of creative thinking. Certainly all the evidence goes to show that no mind is capable of its best efforts when influenced by even small quantities of alcohol. If any reader of these words is disposed to challenge this statement, on the strength of his own personal experience, I would ask him to reflect carefully as to whether what he has been disposed to regard as a stimulant effect may not be better explained along lines suggested by these words of Professor James: "The reason for craving alcohol is that it is an anesthetic even in moderate quantities. It obliterates a part of the field of consciousness and abolishes collateral trains of thought."

The experimental evidence that tends to establish the position of alcohol as an inhibitor and disturber rather than a promoter of mental activity has been gathered largely by German investigators. Many of their experiments are of a rather technical character, aiming to test the basal operations of the mind. Others, however, are eminently practical, as we shall see. The earliest experiments, made by Exner in Vienna so long ago as 1873, aimed to determine the effect of alcohol upon the so-called reaction-time. The subject of the experiment sits at a table, with his finger upon a telegraph key. At a given signal — say a flash of light — he releases the key. The time that elapses between signal and response — measured electrically in fractions of a second — is called the simple or direct reaction-time. This varies for different individuals, but is relatively constant, under given conditions, for the same individual. Exner found, however, that when an individual had imbibed a small quantity of alcohol, his reaction-time was lengthened, though the subject believed himself to be responding more promptly than before.

These highly suggestive experiments attracted no very great amount of attention at the time. Some years later, however, they were repeated by several investigators, including Dietl, Vintschgau, and in particluar Kraepelin and his pupils. It was then discovered that, in the case of a robust young man, if the quantity of alcohol ingested was very small, and the tests were made immediately, the direct reaction-time was not lengthened, but appreciably shortened instead. If, however, the quantity of alcohol was increased, or if the experiments were made at a considerable interval of time after its ingestion, the reaction-time fell below the normal, as in Exner's experiments.

Subsequent experiments tested mental processes of a somewhat more complicated character. For example, the subject would place, each hand on a telegraph key, at right and left. The signals would then be varied, it being understood that one key or the other would be pressed promptly accordingly as a red or a white light appeared. It became necessary, therefore, to recognize the color of the light, and to recall which hand was to be moved at that particular signal: in other words, to make a choice not unlike that which a locomotive engineer is required to make when he encounters an unexpected signal light. The tests showed that after the ingestion of a small quantity of alcohol — say a glass of beer — there was a marked disturbance of the mental processes involved in this reaction. On the average, the keys were released more rapidly than before the alcohol was taken, but the wrong key was much more frequently released than under normal circumstances. Speed was attained at the cost of correct judgment. Thus, as Dr. Stier remarks, the experiment shows the elements of two of the most significant and persistent effects of alcohol, namely, the vitiating of mental processes and the increased tendency to hasty or uncoordinated movements. Stated otherwise, a levelling down process is involved, whereby the higher function is dulled, the lower function accentuated.

Equally suggestive are the results of some experiments devised by Ach and Maljarewski to test the effects of alcohol upon the perception and comprehension of printed symbols. The subject was required to read aloud a continuous series of letters or meaningless syllables or short words, as viewed through a small slit in a revolving cylinder. It was found that after taking a small quantity of alcohol, the subject was noticeably less able to read correctly. His capacity to repeat, after a short interval, a number of letters correctly read, was also much impaired. He made more omissions than before, and tended to substitute words and syllables for those actually seen. It is especially noteworthy that the largest number of mistakes were made in the reading of meaningless syllables,— that is to say, in the part of the task calling for the highest or most complicated type of mental activity.

Another striking illustration of the tendency of alcohol to impair the higher mental processes was given by some experiments instituted by Kraepelin to test the association of ideas. In these experiments, a word is pronounced, and the subject is required to pronounce the first word that suggests itself in response. Some very interesting secrets of the subconscious personality are revealed thereby, as was shown, for example, in a series of experiments conducted last year at Zürich by Dr. Frederick

Peterson of New York. But I cannot dwell on these here. Suffice it for our purpose that the possible responses are of two general types. The suggested word being, let us say, "book," the subject may (1) think of some word associated logically with the idea of a book, such as "read" or "leaves"; or he may (2) think of some word associated merely through similarity of sound, such as "cook" or "shook." In a large series of tests, any given individual tends to show a tolerably uniform proportion between the two types of association; and this ratio is in a sense explicative of his type of mind. Generally speaking, the higher the intelligence, the higher will be the ratio of logical to merely rhymed associations. Moreover, the same individual will exhibit more associations of the logical type when his mind is fresh than when it is exhausted, as after a hard day's work.

In Kraepelin's experiments it appeared that even the smallest quantity of alcohol had virtually the effect of fatiguing the mind of the subject, so that the number of his rhymed responses rose far above the normal. That is to say, the lower form of association of ideas was accentuated, at the expense of the higher. In effect, the particular mind experimented upon was always brought for the time being to a lower level by the alcohol.

The Effect of a Bottle of Wine a Day

When a single dose of alcohol is administered, its effects gradually disappear, as a matter of course. But they are far more persistent than might be supposed. Some experiments conducted by Fürer are illuminative as to this. He tested a person for several days, at a given hour, as to reaction-time, the association of ideas, the capacity to memorize, and facility in adding. The subject was then allowed to drink two litres of beer in the course of a day. No intoxicating effects whatever were to be discovered by ordinary methods. The psychological tests, however, showed marked disturbance of all the reactions, a diminished capacity to memorize, decreased facility in adding, etc., not merely on the day when the alcohol was taken, but on succeeding days as well. Not until the third day was there a gradual restoration to complete normality; although the subject himself — and this should be particularly noted — felt absolutely fresh and free from after-effects of alcohol on the day following that on which the beer was taken.

Similarly Rüdin found the effects of a single dose of alcohol to persist, as regards some forms of mental disturbance, for twelve hours, for other forms twenty-four hours, and for yet others thirty-six hours and more. But Rüdin's experiments bring out another aspect of the subject, which no one who considers the alcohol question in any of its phases should overlook: the fact, namely, that individuals differ greatly in their response to a given quantity of the drug. Thus, of four healthy young students who formed the subjects of Rüdin's experiment, two showed very marked disturbance of the mental functions for more than forty-eight hours, whereas the third was influenced for a shorter time, and the fourth was scarcely affected at all. The student who was least affected was not, as might be supposed, one who had been accustomed to take alcoholics habitually, but, on the contrary, one who for six years had been a total abstainer.

Noting thus that the effects of a single dose of alcohol may persist for two or three days, one is led to inquire what the result will be if the dose is repeated day after day. Will there then be a cumulative effect, or will the system become tolerant of the drug and hence unresponsive? Some experiments of Smith, and others of Kürz and Kraepelin have been directed toward the solution of this all-important question. The results of the experiments show a piling up of the disturbing effects of the alcohol. Kürz and Kraepelin estimate that after giving eighty grams per day to an individual for twelve successive days, the working capacity of that individual's mind was lessened by from twenty-five to forty per cent. Smith found an impairment of the power to add, after twelve days, amounting to forty per cent.; the power to memorize was reduced by about seventy per cent.

Forty to eighty grams of alcohol, the amounts used in producing these astounding results, is no more than the quantity contained in one to two litres of beer or in a half-bottle to a bottle of ordinary wine. Professor Aschaffenburg, commenting on these experiments, points the obvious moral that the so-called moderate drinker, who consumes his bottle of wine as a matter of course each day with his dinner — and who doubtless would declare that he is never under the influence of liquor — is in reality never actually sober from one week's end to another. Neither in bodily nor in mental activity is he ever up to what should be his normal level.

That this fair inference from laboratory experiments may be demonstrated in a thoroughly practical field, has been shown by Professor Aschaffenburg himself, through a series of tests made on four professional typesetters. The tests were made with all the rigor of the psychological laboratory (the experimenter is a former pupil of Kraepelin), but they were conducted in a printing office, where the subjects worked at their ordinary desks, and in precisely the ordinary way, except that the copy from which the type was set was always printed, to secure perfect uniformity. The author summarizes the results of the experiment as follows:

A Loss of Ten Per Cent, in Working Efficiency

"The experiment extended over four days. The first and third days were observed as normal days, no alcohol being given. On the second and fourth days each worker received thirty-five grams (a little more than one ounce) of alcohol, in the form of Greek wine. A comparison of the results of work on normal and on alcoholic days showed, in the case of one of the workers, no difference. But the remaining three showed greater or less retardation of work, amounting in the most pronounced case to almost fourteen per cent. As typesetting is paid for by measure, such a worker would actually earn ten per cent. less on days when he consumed even this small quantity of alcohol."

In the light of such observations, a glass of beer or even the cheapest bottle of wine is seen to be an expensive luxury. To forfeit ten per cent, of one's working efficiency is no trifling matter in these days of strenuous competition. Perhaps it should be noted that the subjects of the experiment were all men habituated to the use of liquor, one of them being accustomed to take four glasses of beer each week day, and eight or ten on Sundays. This heaviest drinker was the one whose work was most influenced in the experiment just related. The one whose work was least influenced was the only one of the four who did not habitually drink beer every day; and he drank regularly on Sundays. It goes without saying that all abstained from beer during the experiment. We may note, further, that all the men admitted that they habitually found it more difficult to work on Mondays, after the over-indulgence of Sunday, than on other days, and that they made more mistakes on that day. Aside from that, however, the men were by no means disposed to admit, before the experiment, that their habitual use of beer interfered with their work. That it really did so could not well be doubted after the experiment.

The Effect of Beer-drinking on German School-children

Some doubly significant observations as to the practical effects of beer and wine in dulling the. faculties were made by Bayer, who investigated the habits of 591 children in a public school in Vienna. These pupils were ranked by their teachers into three groups, denoting progress as "good," "fair," or "poor" respectively. Bayer found, on investigation, that

134 of these pupils took no alcoholic drink; that 164 drank alcoholics very seldom; but that 219 drank beer or wine once daily; 71 drank it twice daily; and three drank it with every meal. Of the total abstainers, 42 per cent. ranked in the school as "good," 49 per cent. as "fair," and 9 per cent. as "poor." Of the occasional drinkers, 34 per cent. ranked as "good," 57 per cent. as "fair," and 9 per cent. as " poor." Of the daily drinkers, 28 per cent. ranked as "good," 58 per cent. as "fair," and 14 per cent. as "poor." Those who drank twice daily ranked 25 per cent. "good," 58% *[sic!]* "fair," and 18 per cent. "poor." Of the three who drank thrice daily, one ranked as "fair," and the other two as "poor." Statistics of this sort are rather tiresome; but these will repay a moment's examination. As Aschaffenburg, from whom I quote them, remarks, detailed comment is superfluous: the figures speak for themselves.

Neither in England nor America, fortunately, would it be possible to gather statistics comparable to these as to the effects of alcohol on growing children; for the Anglo-Saxon does not believe in alcohol for the child, whatever his view as to its utility for adults. The effects of alcohol upon the growing organism have, however, been studied here with the aid of subjects drawn from lower orders of the animal kingdom. Professor C. F. Hodge, of Clark University, gave alcohol to two kittens, with very striking results. "In beginning the experiment," he says, "it was remarkable how quickly and completely all the higher psychic characteristics of both the kittens dropped out. Playfulness, purring, cleanliness and care of coat, interest in mice, fear of dogs, while normally developed before the experiment began, all disappeared so suddenly that it could hardly be explained otherwise than as a direct influence of the alcohol upon the higher centers of the brain. The kittens simply ate and slept, and could scarcely have been less active had the greater part of their cerebral hemisphere been removed by the knife."

The Development of Fear in Alcoholized Dogs

Professor Hodge's experiments extended also to dogs. He found that the alcoholized dogs in his kennel were lacking in spontaneous activity and in alertness in retrieving a ball. These defects must be in part explained by lack of cerebral energy, in part by weakening of the muscular system. Various other symptoms were presented that showed the lowered tone of the entire organism under the influence of alcohol; but perhaps the most interesting phenomenon was the development of extreme timidity on the part of all the alcoholized dogs. The least thing out of the ordinary caused them to exhibit fear, while their kennel companions exhibited only curiosity or interest. "Whistles and bells, in the distance, never ceased to throw them into a panic in which they howled and yelped while the normal dogs simply barked." One of the dogs even had "paroxysms of causeless fear with some evidence of hallucination. He would apparently start at some imaginary object, and go into fits of howling."

The characteristic timidity of the alcoholized dogs did not altogether disappear even when they no longer received alcohol in their diet. Timidity had become with them a "habit of life." As Professor Hodge suggests, we are here apparently dealing with "one of the profound physiological causes of fear, having wide application to its phenomena in man. Fear is commonly recognized as a characteristic feature in alcoholic insanity, and delirium tremens is the most terrible form of fear psychosis known." The development of the same psychosis, in a modified degree, through the continued use of small quantities of alcohol, emphasizes the causal relation between the use of alcohol and the genesis of timidity. It shows how pathetically mistaken is the popular notion that alcohol inspires courage; and, to anyone who clearly appreciates the share courage plays in the battle of life, it suggests yet another lamentable way in which alcohol handicaps its devotees.

Is Alcohol A Poison?

It is perhaps hardly necessary to cite further experiments directly showing the depressing effects of alcohol, even in small quantities, upon the mental activities. Whoever examines the evidence in its entirety will scarcely avoid the conclusion reached by Smith, as the result of his experiments already referred to, which

Dr. Abel summarizes thus: "One half to one bottle of wine, or two to four glasses of beer a day, not only counteract the beneficial effects of 'practice' in any given occupation, but also depress every form of intellectual activity; therefore every man, who, according to his own notions, is only a moderate drinker places himself by this indulgence on a lower intellectual level and opposes the full and complete utilization of his intellectual powers." I content myself with repeating that, to the thoughtful man, the beer and the wine must seem dear at such a price.

To anyone who may reply that he is willing to pay this price for the sake of the pleasurable emotions and passions that are sometimes permitted to hold sway in the absence of those higher faculties of reason which alcohol tends to banish, I would suggest that there is still another aspect of the account which we have not as yet examined. We have seen that alcohol may be a potent disturber of the functions of digestion, of muscular activity, and of mental energizing. But we have spoken all along of function and not of structure. We have not even raised a question as to what might be the tangible effects of this disturber of functions upon the physical organism through which these functions are manifested. We must complete our inquiry by asking whether alcohol, in disturbing digestion, may not leave its mark upon the digestive apparatus; whether in disturbing the circulation it may not put its stamp upon heart and blood vessels; whether in disturbing the mind it may not leave some indelible record on the tissues of the brain.

Stated otherwise, the question is this: Is alcohol a poison to the animal organism ? A poison being, in the ordinary acceptance of the word, an agent that may injuriously affect the tissues of the body, and tend to shorten life.

Students of pathology answer this question with no uncertain voice. The matter is presented in a nutshell by the Professor of Pathology at Johns Hopkins University, Dr. William H. Welch, when he says: "Alcohol in sufficient quantities is a poison to all living organisms, both animal and vegetable." To that unequivocal pronouncement there is, I believe, no dissenting voice, except that a word-quibble was at one time raised over the claim that alcohol in exceedingly small doses might be harmless. The obvious answer is that the same thing is true of any and every poison whatsoever. Arsenic and strychnine, in appropriate doses, are recognized by all physicians as admirable tonics; but no one argues in consequence that they are not virulent poisons.

Open any work on the practice of medicine quite at random, and whether you chance to read of diseased stomach or heart or blood-vessels or liver or kidneys or muscles or connective tissues or nerves or brain — it is all one: in any case you will learn that alcohol may be an active factor in the causation, and a retarding factor in the cure, of some, at least, of the important diseases of the organ or set of organs about which you are reading. You will rise with the conviction that alcohol is not merely a poison, but the most subtle, the most far-reaching, and, judged by its ultimate effects, incomparably the most virulent of all poisons.

Alcohol and Disease

Here are a few corroborative facts, stated baldly, almost at random: Rauber found that a ten per cent. solution of alcohol "acted as a definite protoplasmic poison to all forms of cell life with which he experimented — including the hydra, tapeworms, earthworms, leeches, crayfish, various species of fish, Mexican axolotl, and mammals, including the human subject." Berkely found, in four rabbits out of five in which he had induced chronic alcohol poisoning, fatty degeneration of the heart muscle. This condition, he says, "seems to be present in all animals subject to a continual administration of alcohol in which sufficient time between the doses is not allowed for complete elimination." Cowan finds that alcoholic cases "bear acute diseases badly, failure of the heart always ensuing at an earlier period than one would anticipate." Bollinger found the beer-drinkers of Munich so subject to hypertrophied or dilated hearts as to justify Liebe in declaring that "one man in sixteen in Munich drinks himself to death."

Dr. Sims Woodhead, Professor of Pathology in the University of Cambridge, says of the effect of alcohol on the heart: "In addition to the fatty degeneration of the heart that is so frequently met with in chronic alcoholics, there appears in some cases to be an increase of fibrous tissue between the muscle fibers, accompanied by wasting of these tissues. . . . Heart failure, one of the most frequent causes of death in people of adult and advanced years, is often due to fatty degeneration, and a patient who suffers from alcoholic degeneration necessarily runs a much greater risk of heart failure during the course of acute fevers or from overwork, exhaustion, and an overloaded stomach, and the like, than does the man with a strong, healthy heart unaffected by alcohol or similar poisons."

It must be obvious that these words give a clue to the agency of alcohol in shortening the lives of tens of thousands of persons with whose decease the name of alcohol is never associated in the minds of their friends or in the death certificates.

Dr. Woodhead has this to say about the blood-vessels: "In chronic alcoholism in which the poison is acting continuously, over a long period, a peculiar fibrous condition of the vessels is met with; this, apparently, is the result of a slight irritation of the connective tissue of the walls of these vessels. The wall of the vessel may become thickened throughout its whole extent or irregularly, and the muscular coat may waste away as a new fibrous or scar-like tissue is formed. The wasting muscles may undergo fatty degeneration, and, in these, lime salts may be deposited; the rigid, brittle, so-called pipe-stem vessels are the result." Referring to these degenerated arteries, Dr. Welch says: "In this way alcoholic excess may stand in a causative relation to cerebral disorders, such as apoplexy and paralysis, and also the diseases of the heart and kidneys."

From our present standpoint it is particularly worthy of remark that Professor Woodhead states that this calcification of the blood-vessels is likely to occur in persons who have never been either habitual or occasional drunkards, but who have taken only "what they are pleased to call 'moderate' quantities of alcohol." Similarly, Dr. Welch declares that "alcoholic diseases are certainly not limited to persons recognized as drunkards. Instances have been recorded in increasing number in recent years of the occurrence of diseases of the circulatory, renal, and nervous systems, reasonably or positively attributable to the use of alcoholic liquors, in persons who never became really intoxicated and were regarded by themselves and by others as 'moderate drinkers.'"

"It is well established," adds Dr. Welch, "that the general mortality from diseases of the liver, kidney, heart, blood-vessels, and nervous system is much higher in those following occupations which expose them to the temptation of drinking than in others." Strumpell declares that chronic inflammation of the stomach and bowels is almost exclusively of alcoholic origin; and that when a man in the prime of life dies of certain chronic kidney affections, one may safely infer that he has been a lover of beer and other alcoholic drinks. Similarly, cirrhosis of the liver is universally recognized as being, nine times in ten, of alcoholic origin. The nervous affections of like origin are numerous and important, implicating both brain cells and peripheral fibres.

How the Poison Works

Without going into further details as to the precise changes that alcohol may effect in the various organs of the body, we may note that these pathological changes are everywhere of the same general type. There is an ever-present tendency to destroy the higher form of cells — those that are directly concerned with the vital processes — and to replace them with useless or harmful connective tissue. "Whether this scar tissue formation goes on in the heart, in the kidneys, in the liver, in the blood-vessels, or in the nerves," says Woodhead, "the process is essentially the same, and it must be associated with the accumulation of poisonous or waste products in the lymph spaces through which the nutrient fluids pass to the tissues. The contracting scar tissue of a wound has its exact homologue in the contracting scar tissue that is met with in the liver, in the kidney, and in the brain."

It is not altogether pleasant to think that one's bodily tissues — from the brain to the remotest nerve fibril, from the heart to the minutest arteriole — may perhaps be undergoing day by day such changes as these. Yet that is the possibility which every habitual drinker of alcoholic beverages — "moderate drinker" though he be — must face. This is an added toll that does not appear in the first price of the glass of beer or bottle of wine, but it is a toll that may refuse to be overlooked in the final accounting.

Alcohol and Acute Infections

In connection with experiments in rendering animals and men immune from certain contagious diseases through inoculation with specific serums, Deléarde, working in

Calmette's laboratory in Lille, showed that alcoholized rabbits are not protected by inoculation, as normal ones are, against hydrophobia. Moreover, he reports the case of an intemperate man, bitten by a mad dog, who died notwithstanding anti-rabic treatment, whereas a boy of thirteen, much more severely bitten by the same dog on the same day, recovered under treatment. Deléarde strongly advises any one bitten by a mad dog to abstain from alcohol, not only during the anti-rabic treatment but for some months thereafter, lest the alcohol counteract the effects of the protective serum.

Similar laboratory experiments have been made by Laitenan, who became fully convinced that alcohol increases the susceptibility of animals to splenic fever, tuberculosis, and diphtheria. Dr. A. C. Abbott, of the University of Pennsylvania, made an elaborate series of experiments to test the susceptibility of rabbits to various micro-organisms causing pus-formation and blood poisoning. He found that the normal resistance of rabbits to infection from this source was in most cases "markedly diminished through the influence of alcohol when given daily to a stage of acute intoxication." "It is interesting to note," Dr. Abbott adds, "that the results of inoculation of the alcoholized rabbits with the erysipelas coccus correspond in a way with clinical observations on human beings addicted to the excessive use of alcohol when infected by this organism."

Additional confirmation of the deleterious effects of alcohol in this connection was furnished by the cats and dogs of Professor Hodge's experiments, already referred to. All of these showed peculiar susceptibility to infectious diseases, not only being attacked earlier than their normal companions, but also suffering more severely. This accords with numerous observations on the human subject; for example, with the claim made some years ago by McCleod and Milles that Europeans in Shanghai who used alcohol showed increased susceptibility to Asiatic cholera, and suffered from a more virulent type of the disease. Professor Woodhead points out that many of the foremost authorities now concede the justice of this view, and unreservedly condemn the giving of alcohol, even in medicinal doses, to patients suffering from cholera or from various other acute diseases and intoxications, including diphtheria, tetanus, snake-bite, and pneumonia, as being not merely useless but positively harmful. Even when the patient has advanced far toward recovery from an acute infectious disease, it is held still to be highly unwise to administer alcohol, since this may interfere with the beneficent action of the anti-toxins that have developed in the tissues of the body, and in virtue of which the disease has been overcome.

The Ally of Tuberculosis

Not many physicians, perhaps, will go so far as Dr. Muirhead of Edinburgh, who at one time claimed that he had scarcely known of a death in a case of pneumonia uncomplicated by alcoholism; but almost every physician will admit that he contemplates with increased solicitude every case of pneumonia thus complicated. Equally potent, seemingly, is alcohol in complicating that other ever-menacing lung disease, tuberculosis. Dr. Crothers long ago asserted that inebriety and tuberculosis are practically interconvertible conditions; a view that may be intrepreted in the words of Dr. Dickinson's Baillie Lecture:

"We may conclude, and that confidently, that alcohol promotes tubercle, not because it begets the bacilli, but because it impairs the tissues, and makes them ready to yield to the attacks of the parasites." Dr. Brouardel, at the Congress for the Study of Tuberculosis, in London, was equally emphatic as to the influence of alcohol in preparing the way for tuberculosis, and increasing its virulence; and this view has now become general — curiously reversing the popular impression, once held by the medical profession as well, that alcohol is antagonistic to consumption.

Corroborative evidence of the baleful alliance between alcohol and tuberculosis is furnished by the fact that in France the regions where tuberculosis is most prevalent correspond with those in which the consumption of alcohol is greatest. Where the average annual consumption was 12.5 litres per person, the death rate from consumption was found by Baudron to be 32.8 per thousand. Where alcoholic consumption rose to 35.4 litres, the death rate from consumption increased to 107.8 per thousand. Equally suggestive are facts put forward by Guttstadt in regard to the causes of death in the various callings in Prussia. He found that tuberculosis claimed 160 victims in every thousand deaths of persons over twenty-five years of age. But the number of deaths from this disease per thousand deaths among gymnasium teachers, physicians, and Protestant clergymen, for example, amounted respectively to 126, 113, and 76 only; whereas the numbers rose, for hotel-keepers, to 237, for brewers, to 344, and for waiters, to 556. No doubt several factors complicate the problem here, but one hazards little in suggesting that a difference of habit as to the use of alcohol was the chief determinant in running up the death rate due to tuberculosis from 76 per thousand at one end of the scale to 556 at the other.

Pneumonia and tuberculosis combined account for one-fifth of all deaths in the United States, year by year. In the light of what has just been shown, it would appear that alcohol here has a hand in the carrying off of other untold thousands with whose untimely demise its name is not officially associated. I may add that certain German authorities, including, for example, Dr. Liebe, present evidence — not as yet demonstrative — to show that cancer must also be added to the list of diseases to which alcohol predisposes the organism.

Hereditary Effects of Alcohol

If additional evidence of the all-pervading influence of alcohol is required, it may be found in the thought-compelling fact that the effects are not limited to the individual who imbibes the alcohol, but may be passed on to his descendants. The offspring of alcoholics show impaired vitality of the most deep-seated character. Sometimes this impaired vitality is manifested in the non-viability of the offspring; sometimes in deformity; very frequently in neuroses, which may take the severe forms of chorea, infantile convulsions, epilepsy, or idiocy. In examining into the history of 2554 idiotic, epileptic, hysterical, or weak-minded children in the institution at Bicêtre, France, Bourneville found that over 41 per cent, had alcoholic parents. In more than 9 per cent, of the cases, it was ascertained that one or both parents were under the influence of alcohol at the time of procreation, — a fact of positively terrifying significance, when we reflect how alcohol inflames the passions while subordinating the judgment and the ethical scruples by which these passions are normally held in check. Of similar import are the observations of Bezzola and of Hartmann that a large proportion of the idiots and the criminals in Switzerland were conceived during the season of the year when the customs of the country — "May-fests," etc. — lead to the disproportionate consumption of alcohol.

Experimental evidence of very striking character is furnished by the reproductive histories of Professor Hodge's alcoholized dogs. Of 23 whelps born in four litters to a pair of tipplers, 9 were born dead, 8 were deformed, and only 4 were viable and seemingly normal. Meantime, a pair of normal kennel-companions produced 45 whelps, of which 41 were viable and normal — a percentage of 90.2 against the 17.4 per cent. of viable alcoholics. Professor Hodge points out that these results are strikingly similar to the observations of Demme on the progeny of ten alcoholic as compared with ten normal families of human beings. The ten alcoholic families produced 57 children, of whom 10 were deformed, 6 idiotic, 6 choreic or epileptic, 25 non-viable, and only 10, or 17 per cent. of the whole were normal. The ten normal families produced 61 children, two of whom were deformed, 2 pronounced "backward," though not suffering from disease, and 3 non-viable, leaving 54, or 88.5 per cent., normal.

As I am writing this article, the latest report of the Craig Colony for Epileptics, at Sonyea, New York, chances to come to my desk. Glancing at the tables of statistics, I find that the superintendent, Dr. Spratling, reports a history of alcoholism in the parents of 313 out of 950 recent cases. More than 22 per cent. of these unfortunates are thus suffering from the mistakes of their parents. Nor does this by any means tell the whole story, for the report shows that 577 additional cases — more than 60 per cent. of the whole — suffer from "neuropathic heredity"; which means that their parents were themselves the victims of one or another of those neuroses that are peculiarly heritable, and that unquestionably tell, in a large number of cases, of alcoholic indulgence on the part of. their progenitors. "Even to the third and fourth generation," said the wise Hebrew of old; and the laws of heredity have not changed since then.

I cite the data from this report of the Epileptic Colony, not because its record is in any way exceptional, but because it is absolutely typical. The mental image that it brings up is precisely comparable to that which would arise were we to examine the life histories of the inmates of any institution whatever where dependent or delinquent children are cared for, be it idiot asylum, orphanage, hospital, or reformatory. The same picture, with the same insistent moral, would be before us could we visit a clinic where nervous diseases are treated; or — turning to the other end of the social scale — could we sit in the office of a fashionable specialist in nervous diseases and behold the succession of neurotics, epileptics, paralytics, and degenerates that come day by day under his observation. It is this picture, along with others which the preceding pages may in some measure have suggested, that comes to mind and will not readily be banished when one hears advocated "on physiological grounds" the regular use of alcoholic drinks, "in moderation." A vast number of the misguided individuals who were responsible for all this misery never did use alcohol except in what they believed to be strict "moderation"; and of those that did use it to excess, there were few indeed who could not have restricted their use of alcohol to moderate quantities, or have abandoned its use altogether, had not the drug itself made them its slaves by depriving them of all power of choice. Few men indeed are voluntary inebriates.

Alcohol and the "Moderate" Drinker

It does not fall within the scope of my present purpose to dwell upon the familiar aspect of the effects of alcohol suggested by the last sentence. It requires no scientific experiments to prove that one of the subtlest effects of this many-sided drug is to produce a craving for itself, while weakening the will that could resist that craving. But beyond noting that this is precisely in line with what we have everywhere seen to be the typical effect of alcohol — the weakening of higher functions and faculties, with corresponding exaggeration of lower ones — I shall not comment here upon this all too familiar phase of the alcohol problem. Throughout this paper I have had in mind the hidden cumulative effects of relatively small quantities of alcohol rather than the patent effects of excessive indulgence. I have had in mind the voluntary "social" drinker, rather than the drunkard. I have wished to raise a question in the mind of each and every habitual user of alcohol in "moderation" who chances to read this article, as to whether he is acting wisely in using alcohol habitually in any quantity whatever.

If in reply the reader shall say: "There is some quantity of alcohol that constitutes actual moderation; some quantity that will give me pleasure and yet not menace me with these evils," I answer thus:

Conceivably that is true, though it is not proved. But in any event, no man can tell you what the safe quantity is — if safe quantity there be — in any individual case. We have seen how widely individuals differ in susceptibility. In the laboratory some animals are killed by doses that seem harmless to their companions. These are matters of temperament that as yet elude explanation. But this much I can predict with confidence: whatever the "safe" quantity of alcohol for you to take, you will unquestionably at times exceed it. In a tolerably wide experience of men of many nations, 1 have never known an habitual drinker who did not sometimes take more alcohol than even the most liberal scientific estimate could claim as harmless. Therefore I believe that you must do the same.

So I am bound to believe, on the evidence, that if you take alcohol habitually, in any quantity whatever, it is to some extent a menace to you. I am bound to believe, in the light of what science has revealed: (1) that you are tangibly threatening the physical structures of your stomach, your liver, your kidneys, your heart, your blood-vessels, your nerves, your brain; (2) that you are unequivocally decreasing your capacity for work in any field, be it physical, intellectual, or artistic; (3) that you are in some measure lowering the grade of your mind, dulling your higher esthetic sense, and taking the finer edge off your morals; (4) that you are distinctly lessening your chances of maintaining health and attaining longevity; and (5) that you may be entailing upon your descendants yet unborn a bond of incalculable misery.

Such, I am bound to believe, is the probable cost of your "moderate" indulgence in alcoholic beverages. Part of that cost you must pay in person; the balance will be the heritage of future generations. As a mere business proposition: Is your glass of beer, your bottle of wine, your high-ball, or your cocktail worth such a price?

Alcohol Trance[3]

I propose to describe in a general way a peculiar mental state following the toxic use of alcohol, which has only recently attracted attention, and which promises to be a very imporant factor in the medical jurisprudence of the future. Morbid states of the nervous system, in which the mind seems to act automatically, and without consciousness of the surroundings, and with no registration by the memory of these acts, are not new to students of mental and nervous diseases; but the fact that they are more or less common in inebriety from alcohol, and may follow any excess, is a recent discovery. In 1879 I published a short paper "On Trance and

---

[3] By T. D. Crothers

Loss of Consciousness following Inebriety," which, as far as I can ascertain, was the earliest study of these cases ever made. The following are among the first cases which attracted my attention to this subject. In 1877 a patient was admitted to the asylum at Binghamton, with this incident in his history: A year before, while apparently sober, he purchased a trotting-horse, paying a fabulous price. Two days after, he denied all knowledge of the transaction, and became involved in a lawsuit. On the trial it appeared that the purchase of the horse had been discussed for many hours, and that the buyer had exhibited great sagacity and judgment to avoid deception; also that, although drinking large quantities of spirits, he gave no evidence of other than good judgment, and perfect knowledge of his acts and their consequences. In the defense it was shown that the purchase of the horse was a most unusual act; that he never showed any interest in fast horses, or racing, nor had he been on the race-course, and was in fact afraid of driving fast horses; and, lastly, he had many horses in his stables, and needed the money paid for this horse, for a distinct purpose, which had been determined on before. From his own testimony he had many blanks of memory while drinking, and at this time had lost all recollection of passing events from the hour of dinner, during which he drank freely, until next morning, when he drank again and fell into another blank which lasted thirty-six hours. Other testimony indicated a gradual increasing dullness and abstractedness of manner during this time; also apparent disinclination to fix his attention on any one thing long. The suit went against him, and he soon after was brought to the asylum. In another case the president of a bank, a man of wealth and irreproachable character, forged a large check, put the money in his pocket, and the day after was amazed to find the money and to account for it. In an investigation it was proved that he suffered from these blanks of memory after drinking wine freely; that he had before done many unaccountable acts, apparently fully conscious at the time, and yet afterward disclaimed all memory of them, a fact which was supported by their motiveless character. This mental condition may be described as a loss of memory and consciousness of present and passing events, that is concealed and not apparent from a general study of the conduct; or, in other words, a state of the brain similar to somnambulism in respect to the unconscious character of the acts, and all recollection of them. For the time being the sufferer is a literal automaton, giving little or no evidence of hisactual condition, and acting from impulses unknown, and motives that leave no trace.

The late Dr. Beard believed this state to be one of general lowered brain-function, in which the cerebral activity is concentrated in some limited region of the brain, and is largely suspended in the rest. He also urged that the plane of consciousness was below the point of remembering; hence these cases were conscious at the time, but the memory failed to record the impression. In confirmation of this, the late Dr. Forbes Winslow recorded a case of a somnambulist who, while walking about, set his night-dress on fire, and with excellent judgment and coolness threw himself on the bed and extinguished the flames, then resumed his walk, and awoke next morning with no memory of it, and was greatly alarmed at the charred appearance of his dress. Whatever the pathology may be, it is clear that this is a state of irresponsibility, and for the time being a form of dementia and insanity, about which there can be no question. Careful study of those cases for many years has indicated the startling fact that they are very common in inebriety; also that in every case where alcohol is used to excess there are histories of loss of memory and consciousness of acts committed while using spirits. These conditions are almost infinite in variety and complexity, and are considered mere freaks of memory by many persons. Probably in a majority of cases in the early stages these blanks of consciousness and memory are partial, and appear in the delirium or stupor which follows excess of spirits, or in mental states approaching it, and clear up after recovery, or remain like a cloud for weeks, then from some little circumstances break away and every act is recalled. In other cases only a dim, vague impression remains of what has transpired in the past, which may or may not become clear with time; or the blank may be total for the time being, and then break away. In many of these cases there is apparent realization of all his acts and words, in others a self-evident unconsciousness of them. This is only the beginning of another and more pronounced stage, in which the blank of memory and consciousness is total, and during this period the acts and appearance of the person differ but little from those of usual health. In many cases the brain function or action, as seen in his acts, is fully up to the best state of health, even showing more than usual strength in some directions. In a paper read before the Medico-Legal Society of New York, in 1881, I discussed this condition as a trance state following inebriety; since that time a number of different names have been suggested by authors, such as inebriate automatism, inebriate insanity, inebriate unconsciousness—all describing the same condition. The following may be mentioned as facts that are generally accepted as landmarks from which further study may be dated:

1. This trance state is a common condition in inebriety, where, from some peculiar neurotic state, either induced by alcohol, or existing before alcohol was used, or exploded by this drug, a profound suspension of memory and consciousness and literal paralysis of certain brain-functions follow.

2. This trance state may last from a few moments to several days, during which the person may appear and act rationally, and yet be actually a mere automaton, without consciousness or memory of his actual condition.

3. This trance state may be noted by criminal impulses and by unusual thoughts and acts

foreign to all the man's past history. In all these cases there are no apparent disturbances of the nervous system, no convulsions, no premonitions to mark this state; at some unknown point, all unconscious, the eclipse begins, A comparison of the history of a number of cases will show three mental conditions quite prominent: 1. In which the mind in this state acts along certain accustomed lines of thought and action; 2. In which the mind displays unusual ranges of thought and action, which in some cases can be traced to certain mental states growing out of the surroundings; and, 3. Where criminal impulses are prominent, that have no apparent connection with the present or past. These conditions may be illustrated in the following cases: A railroad conductor, who drank to excess every night after the day's work was over, would frequently get up in the morning, go out on his train, perform all his duties correctly, and recover consciousness of himself suddenly on the road, and all the past be a blank to Lira from some point the night before. These blanks occasionally lasted twenty-four hours, and he could never recall anything which happened, and only knew by the money and tickets that he had made a trip on his train. After a time he would put down in a note-book events of importance in this state, which he never did otherwise. The train-hands knew that he was, as they termed it, "memory-drunk," when he used his note-book freely, and seemed dull and abstracted. A pilot on a Sound steamer, after seasons of hard work, and exhaustion from loss of sleep, would use brandy to keep up, and have blanks of hours from which he would recover, having no recollection of what had happened. He would act as usual, only be less talkative, and dull in his manner. A skilled mechanic, who used spirits to excess, suffered from blanks of many hours' duration, during which he attended a dangerous machine, performing all the duties, requiring both skill and judgment. A clergyman, who drinks wine, has frequently conducted service, and preached a sermon without any memory of the fact, having a blank of all surroundings for hours. A grocer, after a period of great excess in the use of spirits, will conduct his business for hours without any consciousness of events, and only know by the books and the statements of others what has taken place. These are only a few of the histories of a large number of cases which I have gathered to illustrate the fact that in this trance state the mind may work along accustomed lines of thought and action. In this condition, the evidence of a mental blank is more or less obscure. In the next division, the mind displays unusual ranges of thought and action, some of which can be traced to the surroundings. A physician, who drank constantly, and was a bitter skeptic, went into a revival meeting and professed change of heart, and took part in the exercises, and the next morning had no recollection of it. Later, while drinking, he heard the singing of the revival meeting, and, dropping all business, entered and took a very active part, and seemed fully conscious of all the surroundings, yet, after a night's sleep, had no recollection whatever of anything which had occurred. In this case the trance state was manifest in unusual deeds and acts, suggested from the surroundings. A similar case was that of an editor, who, after drinking to excess, could always be found in temperance-meetings, making eloquent appeals, and yet he gave no evidence of being under the influence of spirits, nor could he remember anything of what had occurred. Another case is that of a man of fortune, who drank wine freely, awoke and found that he had married his servant, and made an unusual disposition of his property, which was all a blank to him. To his friends and others he seemed fully conscious of the nature and consequences of these events at the time. I think it will be found that inebriates brought suddenly into conditions of excitement are moved by circumstances and surroundings to which they are often really oblivious. If the trance state is present, the influence of the surroundings can not be estimated. The last division, that of *criminal impulse growing out of this trance state,* illustrates the subject of our paper more closely. The following cases bring out the facts better than any description: An inebriate was repeatedly arrested for horse-stealing, and often punished. The crime was committed under similar circumstances, and no. attempt was made to conceal the property; on two occasions he assisted the owner to hunt up the horses. When it was apparent that he was guilty, great was his astonishment, and he denied all recollection of any circumstances or events. This was confirmed by all the circumstances of his life, by his inebriety and blanks of memory, and absence of motive and object in the crime. He was fond of horses, and seemed at this time to be governed by an impulse to drive and ride behind a good horse. A farmer of quiet, good disposition suffered from blanks of memory after drinking to excess. One day, in what seemed full consciousness of the surroundings, he attacked a stranger and injured him so that he died. He had no recollection of the time, purpose, or any circumstances of the tragedy. A periodical drinker, of wealth, tired his buildings, and awaking when they had burned down, offered a large reward for the incendiary. To his great astonishment, the fire was readily traced to him; the circumstances and motive were all a perfect blank. A man of much talent and eminence, who drinks occasionally to excess, has on many occasions offered violence to his wife, whom he loves very dearly. On these occasions he is apparently sober, gives reasons for his conduct, and afterward has not the slightest recollection of it. In a murder-trial recently, it appeared that a drinking man drank early in the morning, then killed his wife, and went about his work in the vicinity, as if nothing had happened, all unconscious until arrested. He was sentenced for life, but has a firm conviction that he did not commit the crime, because he cannot conceive of a motive, and has no recollection of it. A clergyman committed a rape under the most extraordinary circumstances, and denied all recollection of it; his drinking habits and all the incidents of the case sustained his statement. A lawyer of reputation planned the abduction of a lady he was going to marry. A man of a large family and happy domestic relations married a notorious woman. A physician stole a large sum of money from a patient. A college graduate enlisted in the army. In each of these cases there was a history of drinking to excess, and each had no memory of the event, and all the circumstances were so unusual and at variance with previous conduct that undoubtedly a trance state was present. These cases might be multiplied almost indefinitely from the records of criminal courts everywhere. Every day the papers record cases of crime, without motive or purpose, by inebriates who, in defense, claim to have no recollection of it; but, as they were not wildly delirious or stupid at the time of committing the act, they are punished as fully responsible, When the crime is of magnitude, and the defense is insanity, the explanation and theory are so far from the accepted views of experts as to confuse courts and juries, and be criticised and ridiculed by others. This defense occurs most frequently in two forms of cases: One, of a chronic inebriate, who is all the time more or less under the influence of spirits, and who lives in a low moral atmosphere, in bad physical surroundings. Suddenly he commits a crime, which is without motive, and seems a mere accident and result of unforeseen conditions. The second case is of a man who may be a periodical inebriate, and of good character and reputation in everything except excess of use of spirits; whose surroundings and general standing are good, and who commits a homicide or some strange crime under circumstances that are inadequate to explain or account for it. In both of these cases there is no recollection of any of the circumstances, and the defense is based on some specious reasoning and theories. There are evidently disorganized brain-power, mental and physical incoordination, with defect and unsoundness of the reasoning powers, which cannot be made clear to the court and jury. The prevalence of the theological theory, that all these strange, unaccountable acts of inebriates, who are not stupid at the time, or wildly delirious, come from vice and sin, is fatal to all scientific study and progress. This condition of trance, noted by absence of memory and consciousness, has been discussed by Dr. Carpenter, of England, under the title of "Automatic Cerebration," from which I quote the following sentence: "I have noticed some cases of drunkenness, in which a suspension of memory and consciousness was noted, coming on unexpectedly, and then the patient was a victim to morbid impulses which he never realized or had any recollection of after." Dr.

Hughlings Jackson writes at some length on mental automatism, following transient epileptic paroxysms, in which this same condition is described at length as a form of sudden paralysis of the cerebral functions, or conditions of hyperæmia and suspension of some controlling centers. The late Dr. Forbes Winslow describes a similar condition of trance and automatism where the person seemingly acted as fully recognizing right and wrong, although consciousness was obliterated. Dr. Hammond mentioned the case of a man who, after an attack of epilepsy, went about for eight days in a trance state, doing business, and having no memory of it. Dr. Hughes has also mentioned similar cases. Abroad many eminent specialists, including such names as Drs. Bucknill, Clouston, Mercer, and Motet, of Paris, and others, have described this state associated with epilepsy, and following mental shocks in persons who are drunkards. These references are presented to show that the trance state has been observed by eminent men, although not yet studied from the side of crime and responsibility. A large number of cases are constantly before the courts on trial for crime committed after and during excess in the use of alcohol—crime that is purposeless, without motive or object, and differing in the manner of execution, and effort to conceal afterward, from other crime of similar nature—in some cases noted for apparent coolness, without excitement, and cold-heartedness or indifference to the nature of the act. In the defense, all recollection or consciousness of the event is denied, and many circumstances, seen both before and after the crime was committed, bear out this statement. These cases receive no study, and are punished, the result of which precipitates the victim into worse and more degenerate stages. Undoubtedly these cases are suffering from alcoholic trance, and have crossed the border-line of sanity and responsibility, and are as truly insane as the wildest maniac. In this trance state the person is a mere automaton in motion, either moving along certain fixed lines of conduct, or acting in obedience to unknown forces which may change or vary any moment. Some governing center has suspended, and all rememberable consciousness of time and the relation of events has stopped. Changing thoughts and impulses, the suggestion of a disturbed organ, or the impression of a thought or desire felt in the past, may suddenly concentrate into action irrespective of consequences. Both subjective and objective states, influenced by conditions of health and brain-power, may develop into acts that will be unknown and unrecorded by the higher brain-centers. Clinical facts within the observation of any one will indicate, without any kind of doubt, that in all cases of inebriety there are a defective brain-power and ability to recognize the natural relations of life in all particulars. The sufferer is more or less incapable of healthy normal thought and action; he has opened the door for many complex nervous disorders, and the natural process of tearing down the structure is greatly accelerated. If the trance state is found to be present, he has passed into the realm of practical irresponsibility and unconsciousness of the nature and character of his actions. I believe the following summary will be found to outline the future recognition and treatment of these cases:

1. Inebriety in all cases must be regarded as a disease, and the patient forced to use the means for recovery. Like the victim of infectious disease, his personal responsibility is increased,

and the community with him are bound to insist on the treatment as a necessity.

2. Inebriety must be recognized as a condition of legal irresponsibility to a certain extent, depending on the circumstances of each individual case.

3. All unusual acts or crimes committed by inebriates, either in a state of partial stupor or alleged amnesia (or loss of memory), which come under legal recognition, should receive thorough study by competent physicians, before the legal responsibility can be determined.

4. When the trance state is established beyond doubt, the person is both physiologically and legally irresponsible for his acts during this period. But each case should always be determined from the facts of its individual history.

In the light of science the present legal treatment of inebriety is but little else than barbarism. The object of the law, in punishment, benefits no one, and makes the patient more incurable—destroying all possibility of recovery and return to health again. Inebriety in any form may be no excuse for crime in a legal sense, but it is still less an excuse for punishment, which destroys the victim, or makes him more helpless and hopeless. A vast army of inebriates, hovering along these border-lands of disease and crime, who are unknown and unrecognized, except "as vicious and desperately wicked," are a perpetual menace to all progress and civilization, unless they can be reached and checked by rational, effective methods. A revolution of sentiment and practice is demanded, in which the inebriate and the conditions which developed his malady shall be understood; then the means for prevention, restoration, and recovery can be applied along the line of nature's laws